LAND

by Robin Nelson

first step nonfiction

Lerner Publications · Minneapolis

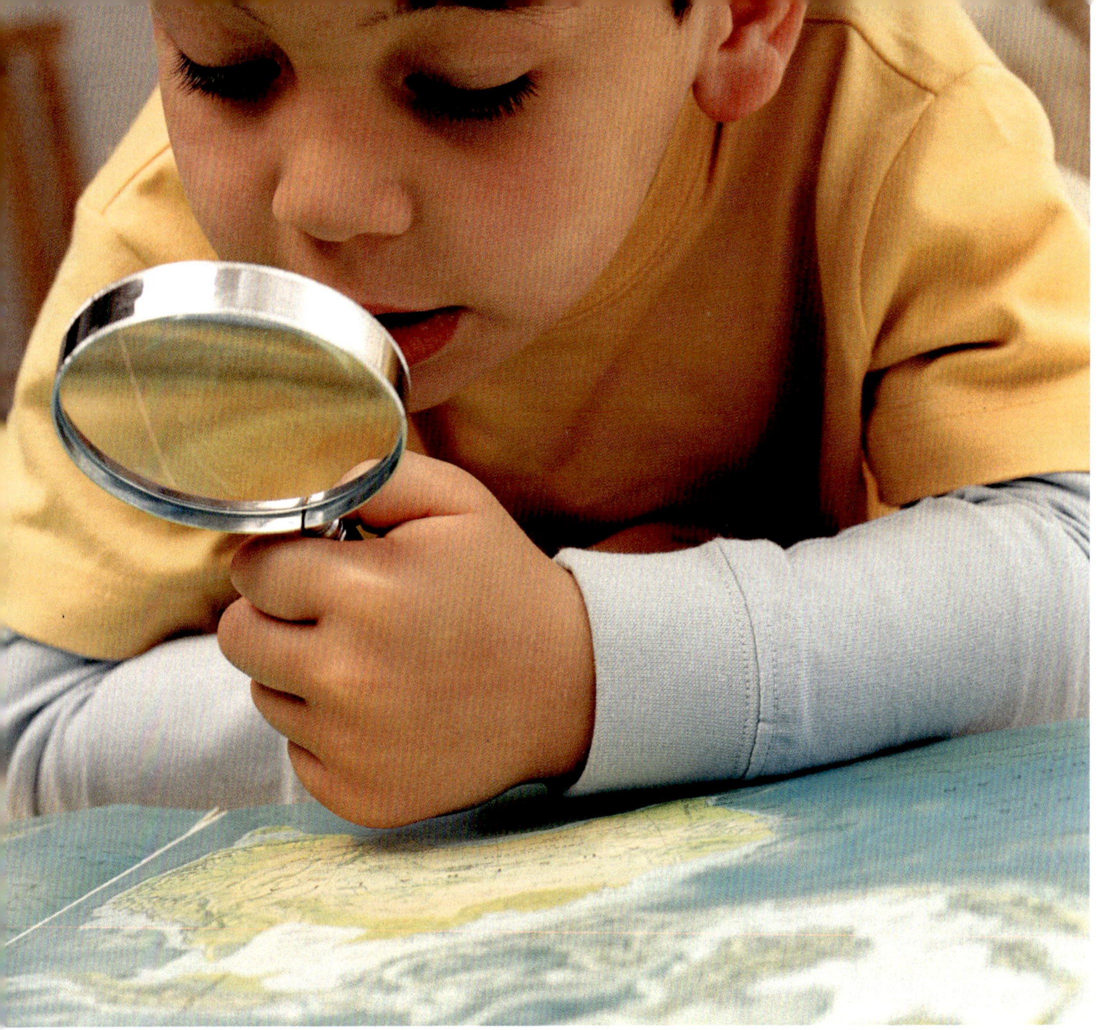

Can you find land?

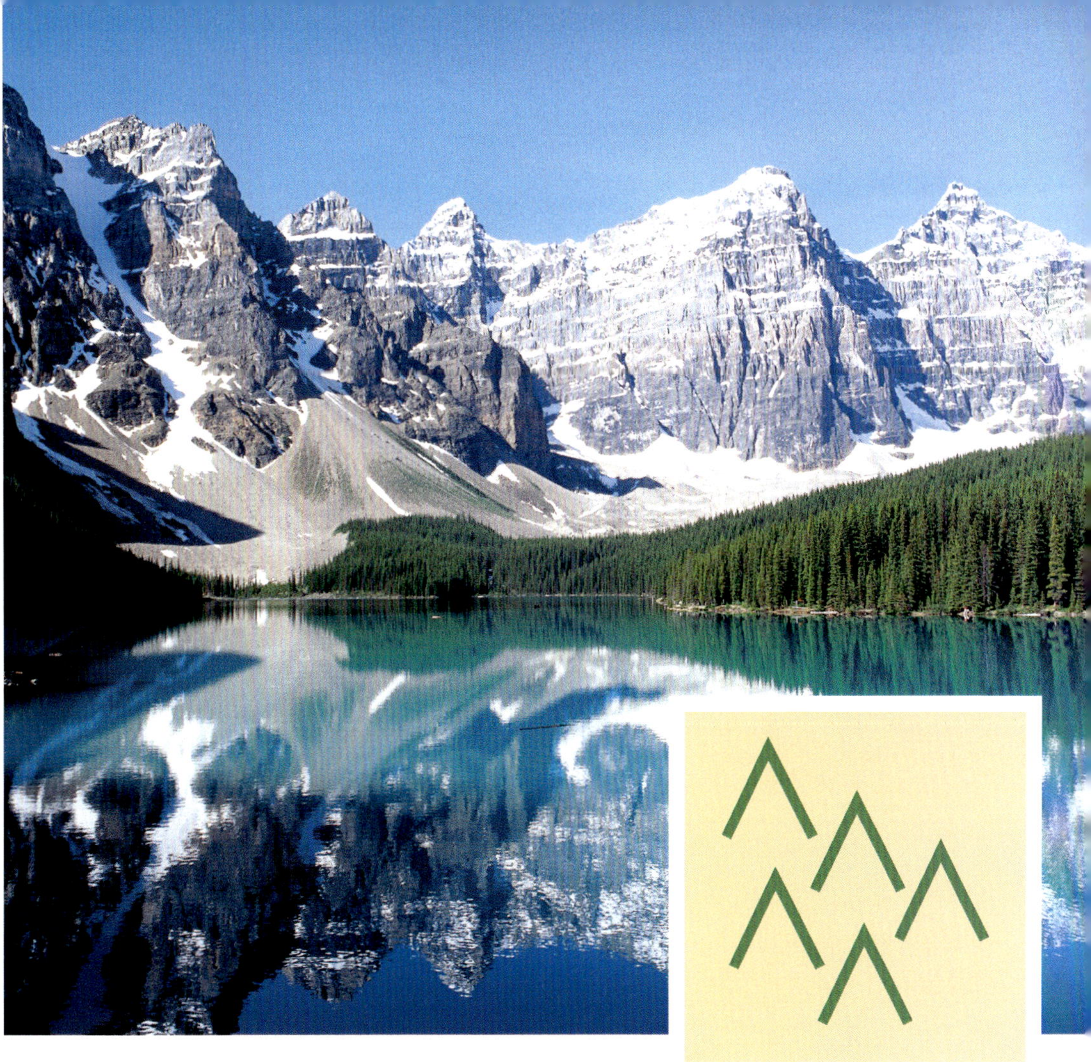

Mountains

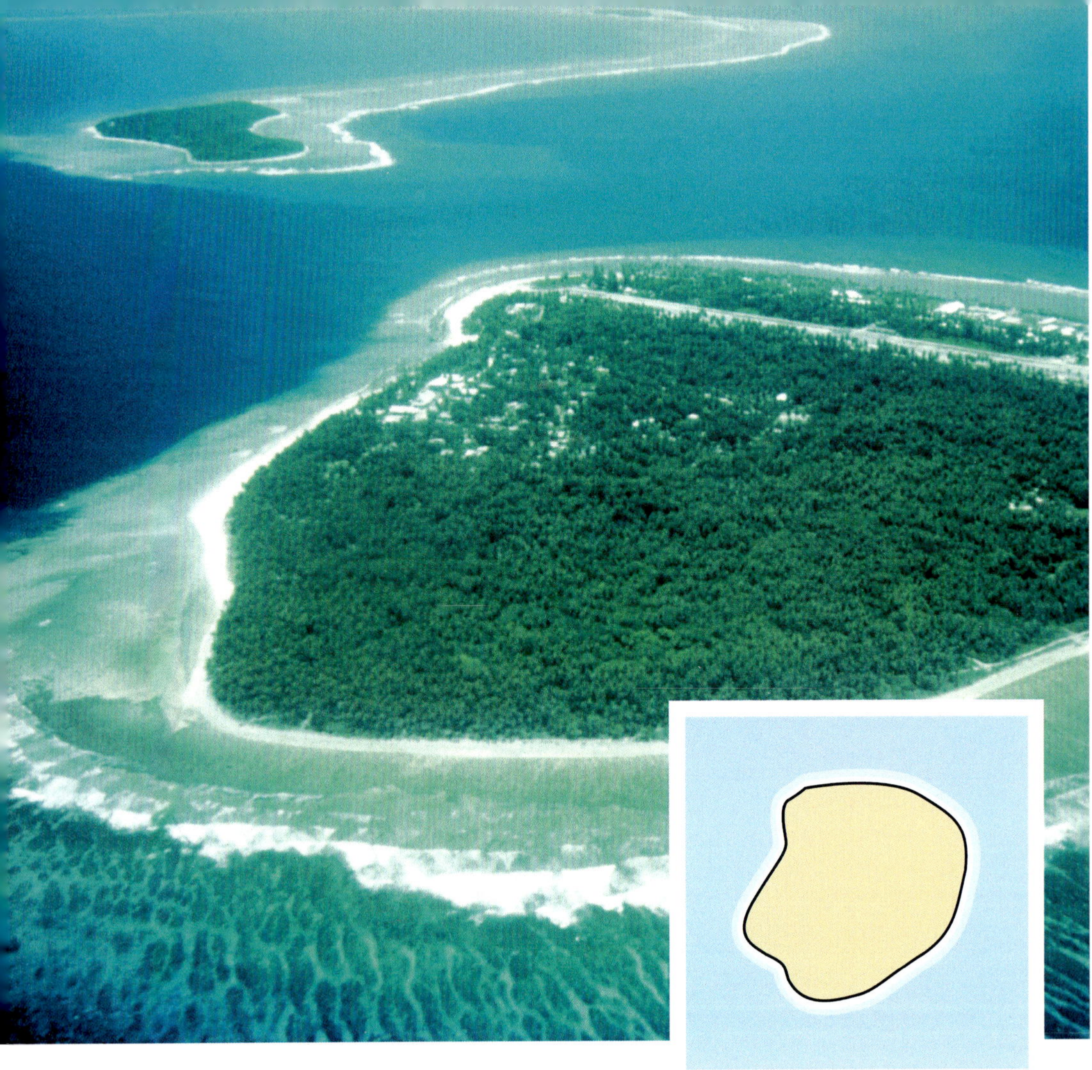

Islands

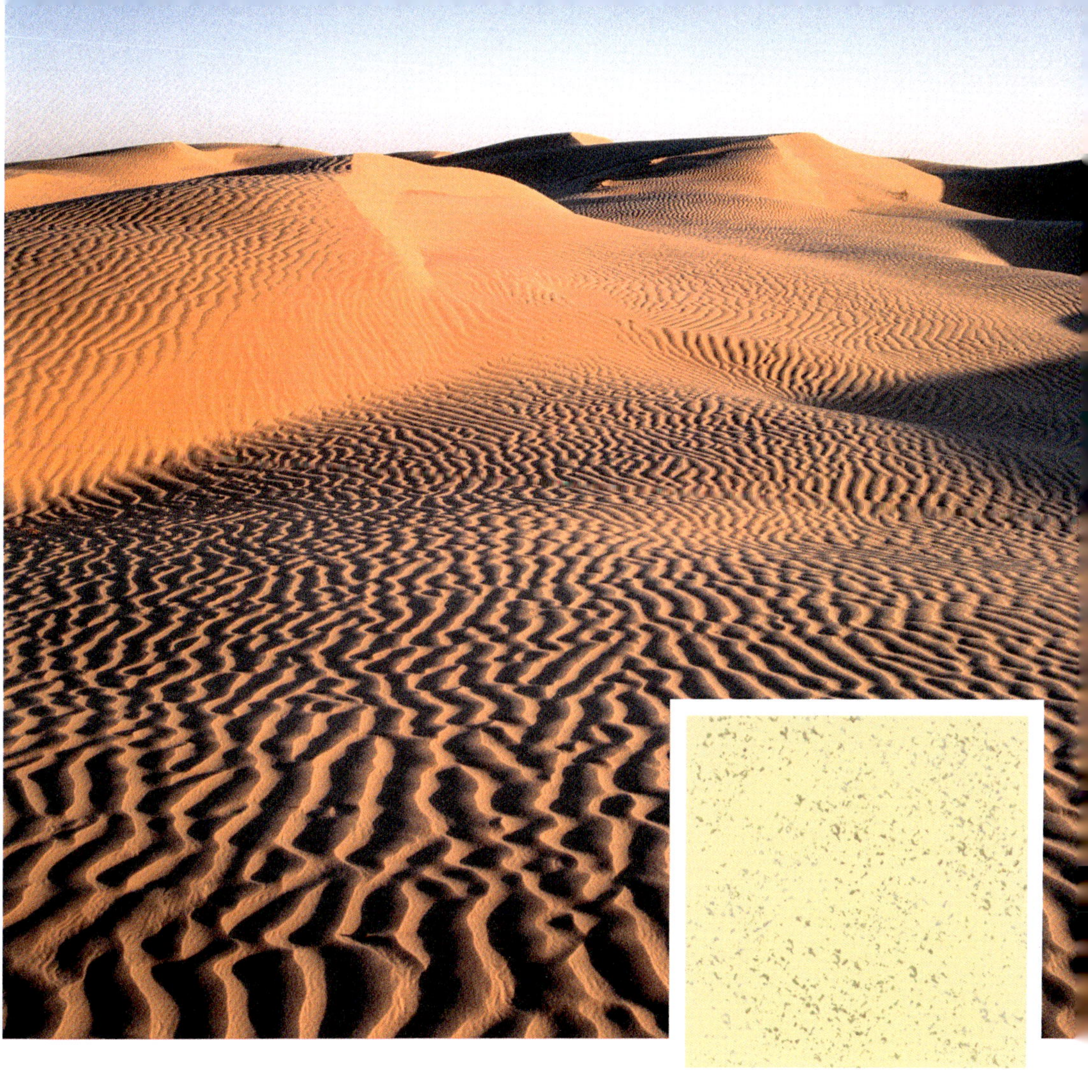

Deserts

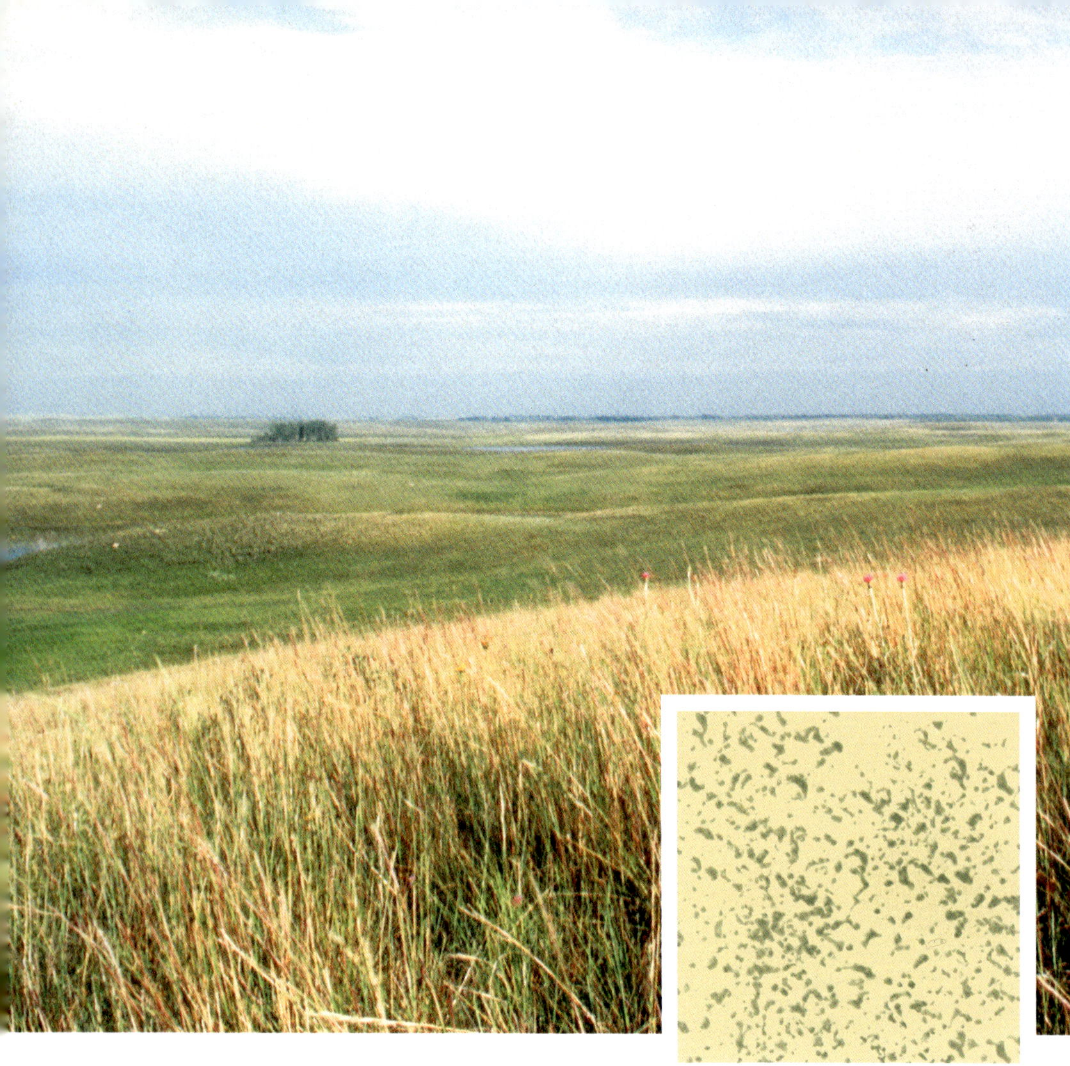

Plains

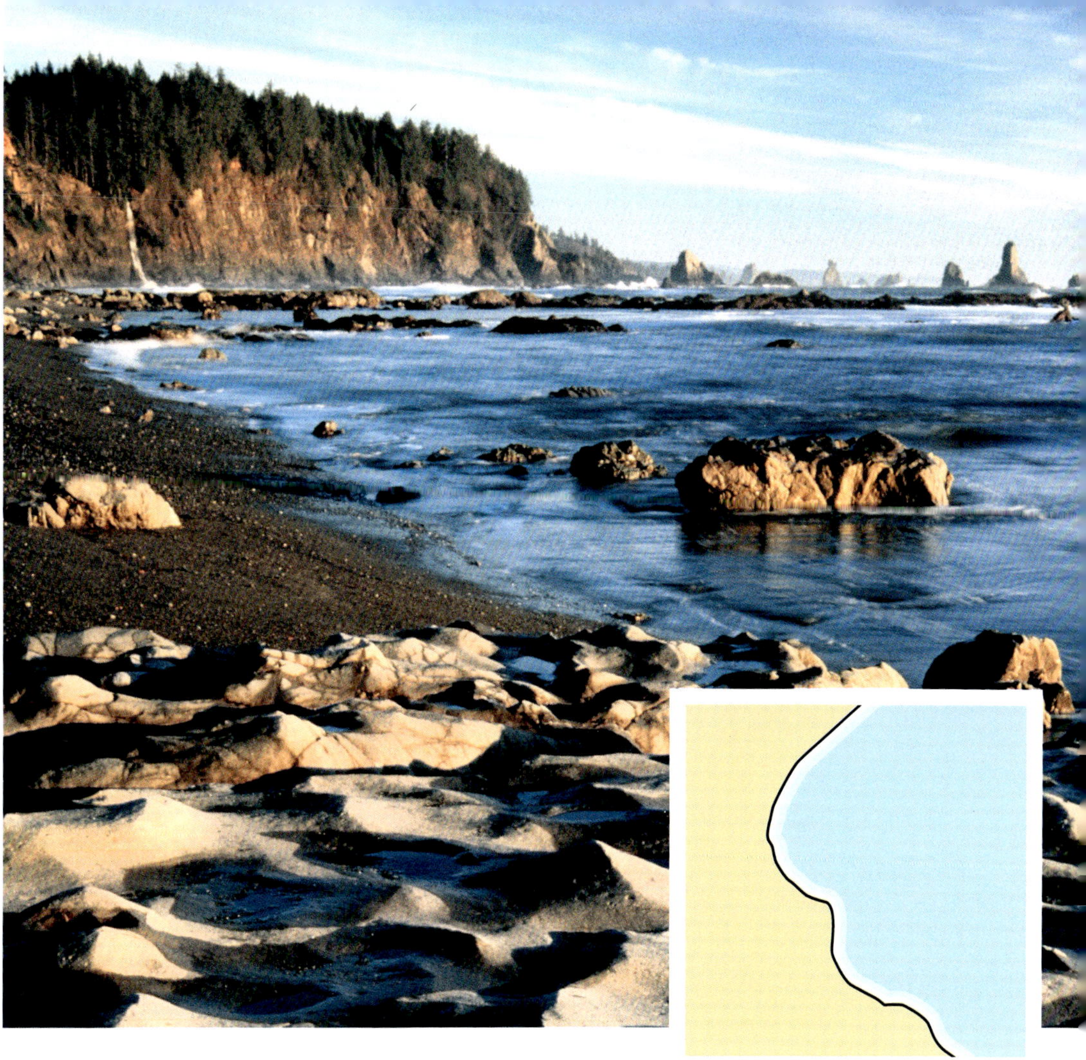

Coasts

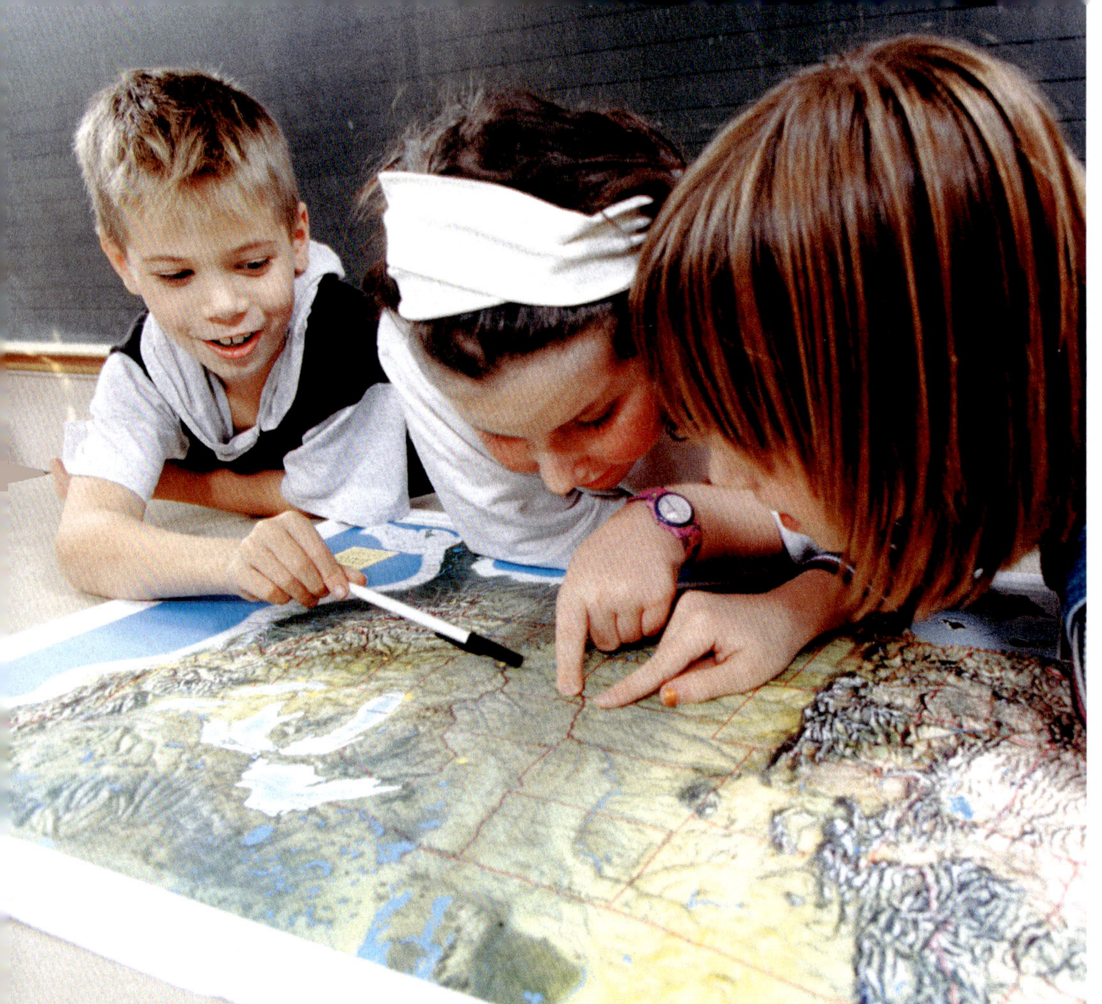

Maps show land.